Focus

7 Ways To Improve Focus And Concentration

Arthur Davis

Copyright © 2013 The Westfield Group LLC

All Rights Reserved. No portion of this book may be reproduced, stored in a retrieval system, or transmitted in any form or by any means – electronic, mechanical, photocopy, recording, or any other – except for brief quotations in printed review, without the prior permission of the publisher.

First Printing, 2013

ISBN-13:978-1493698745
ISBN-10:1493698745

Printed in the United States of America

Disclaimer

This publication is designed to provide accurate and authoritative information in regard to the subject matter covered. It is sold with the understanding that the publisher and author are not engaged in rendering health, legal, counseling, or other professional services. If professional advice or other expert assistance is required, the services of a competent professional person should be sought. Designated trademarks and brands are the property of their respective owners.

DEDICATION

This book is dedicated to all of the health professionals and volunteers, research scientists and many more studying, researching and providing therapy in the field of mental concentration, attention and focus.

CONTENTS

	Acknowledgments	i
	Introduction	1
1	What Is Attention	3
2	Research Into The Process Of Human Attention	7
3	Divided Attention	11
4	The Impact Of Load On Attention Processing	15
5	The Heavy Cost Of Multitasking	19
6	Get To Know How Your Mind Works	23
7	Seven Ways To Improve Your Focus And Concentration	27
	Conclusion	31

ACKNOWLEDGMENTS

I acknowledge and thank all those who made this book possible. My special thanks goes to Mary, my Editor in Chief. Without your editing and input, this book would not be possible.

INTRODUCTION

Think about paying attention and mental focusing the same way you might think about sunshine. Feel the warmth and notice the bright light as rays gently fall to earth from the sun. However when these same rays are highly focused through a magnifying glass, they become powerful enough to ignite a fire or even to cut through something as strong as steel. Think about how a nice rain can "clean" your street but concentrated, it has enough energy to turn an electrical generator or "wash away" whole neighborhoods.

The same goes for human productivity. Someone who is constantly distracted with trivial tweets, computer games, a constant flow of junk email and Smartphone apps will not be nearly as successful as someone who can and does focus his or her mind and behavior. Learning to do this is truly like learning how to unleash the power of your mind. You have a potent tool available when your mind is focused and you make a purposeful decision to concentrate.

Imagine the success, effectiveness and power you can have simply by learning to focus your mind on what's important. What could you achieve if you finish the tasks and solve the problems you need to solve in order to move closer to your coveted goals rather than wasting all day every day on trivia that, at the end of the day, produces nothing and costs everything?

It seems it is harder with each new day for most of us to get and remain focused and to concentrate long enough on one thing to actually get it done. Why is this? Well, there is some new and interesting research that shows there is a cost to pay for the wonders of our information technology revolution. The days of being able to "hole away" somewhere and get things done are gone for many of us. The problem is that the whole world is now able to interrupt us by way of the little beeps and tweets coming from the Smartphone we always have in our hand or pocket. And, we are addicted to these interruptions. If we don't get them several times per hour, we go "surfing" for them in one or several browsers.

We think we are getting more things done faster. We call it multitasking. The problem is that our brains are not able to effectively multitask with complex tasks. In fact, there is some evidence that shows we lose up to 40 percent of our productivity flipping around from one thing to the next. Let me ask you, how many times do you check your email, Facebook page and Twitter account per hour? If you are the average heavy multitasker, your answer will be 37 times plus! Now ask yourself this question. What is this doing for you? Are you smarter? Are you wealthier? How about happier? Maybe it is high time you seriously looked into this whole idea of focused attention. Could be you are one of those heavy multitaskers that needs a "dopamine squirt" your brain gets from your computer or Smartphone every few minutes in order to make it through the day.

1 WHAT IS ATTENTION

The most widely accepted definition of attention goes all the way back to the late eighteen and early nineteen hundreds. During this period a prominent philosophical psychologist by the name of William James defined attention as "the taking possession of the mind, in clear and vivid form, of one out of what seem several simultaneously possible objects or trains of thought".

Today attention is understood to be the ability of any living thing to sense stimuli that surrounds it. Human attention is thought more of as sustained concentration on one specific idea, sensation, thought or activity. This focus of attention enables us to use sensory and brain based information processing systems having limited capacity, to filter vast amounts of information coming from our senses or memory.

Paying Attention

Paying attention requires the selection of some information out of an infinite array of incoming environmental stimuli for additional processing by our brain. Part of the processing our brain does on the information is for the purpose of giving meaning to it. A filtering system at the doorway of our nervous system will allow only certain stimuli meeting particular requirements to go through. The information that does get through the filter is compared to what we know already, so that we can recognize stimuli that we already know and therefore determine what it means to us.

The Two Components Of Attention

There are really two components of attention. One might be called arousal. Information must "arouse" the senses to get noticed. The next component of attention is the selection of the information. The arousal process is one of getting to and staying at a state of alertness sufficient enough to stay "in contact" with the stimuli in the environment. So attention really is what makes the difference between the mental state of being awake and being asleep or even in a coma. When attention is focused, it results from a mental process that selects stimuli and allows it into the conscious so that the brain can decide to respond to the information immediately or store it in memory. The amount of information that is allowed into one's consciousness is only a small amount of the available stimuli at any one moment in time. Someone's capability of being able to switch or "orient" their attention is critical to successfully using attention in different environments.

To Learn Or Store Memories One Must Be Able To Pay Attention

A significant amount of research reveals that learning and overall performance can be significantly degraded in the absence of clear awareness of the stimulus around us. Although attention doesn't seem necessary for at least some minimal level of perceptual processing, it does however seem to be necessary for an experience to enter into one's consciousness or to be stored in memory.

There has been significant research into the anatomy of focused attention. Regarding the physical workings of focused attention, one can think of the process of paying attention much like breathing or blood flow. The mental and physical process of paying attention involves specific body organs. Hearing and vision are critical capabilities that affect one's ability to focus attention. All of these "ingredients" of focused attention can be degraded with injury, aging or even certain mental states such as stress or elation.

Attention Is A Critical Life Skill

Undoubtedly the ability to focus attention is a critical process for lifetime success. Without the ability to filter and process the huge amount of information coming from our senses, we would constantly be in a state of being overwhelmed. Without the ability to focus on specific stimuli in our surroundings, everything would be just a confusing blur. Understanding how humans view the world around them and thus interact and behave in it can only be achieved in the context of understanding how and why humans focus their attention.

Arthur Davis

2 RESEARCH INTO THE PROCESS OF HUMAN ATTENTION

The Early Study Of Attention

Techniques and methodologies for studying attention have evolved and expanded over the years. In the late 1800's research into human attention started out with very simple experiments involving swift responses to either one or a small number of objects in an attempt to determine just how limited people's capacity is for focusing attention on information.

The Stroop Effect

The Stroop Effect is an example of the earliest type of attention testing. The test involves asking subjects to correctly identify the color of ink a word appears in when the word spells out a different color. For example, the word yellow is written in blue ink. This requires someone to ignore what the word means and to focus their attention on the color of the ink the word is written in.

Visual Search Tests

Another method to measure attention is the task of visual search. The early results of these tests were interesting in that they suggested when gazing at a scene of "nature", changes in the scene that were due to changes in brilliance or motion were noticed even as dramatic changes to other content in the scene not in focus were ignored. These results suggest that the old theory of humans being fully aware of the world around them is mostly a fantasy. This shows that people are not aware of most things around them but do have the ability to focus on something that changes.

Attention and Cognition

As computer models were developed in the 1950's, determining how information flows in the nervous system and cognitive processes were simulated. Then in the 1970's, micro electrodes were inserted into the brains of awake and alert monkeys showing that the firing of cells in certain areas of their brain were increased when monkeys paid attention to stimuli that were within the cell's receptive field. This research evolved in the 1980's and 1990's to include the investigation of the entire brain during tasks requiring attention. Studying the entire brain during the attention process improved the older more traditional testing methods and also allowed for brain lesions and EEG (scalp electrode recordings) to be included in the testing. The modern ability of recording changes to the brain over time has improved and validated medicine and psychological therapies to improve a patient's attention span.

Theories Of Visual Attention

There are two main theories in cognitive psychology that attempt to explain how visual attention works. For the most part, visual attention is hypothesized to function in two stages. During the first stage visual attention is distributed evenly over the scene and the information input is processed uniformly. In the next stage attention is focused on a specific visual area and processed serially.

The Spotlight Model

The first model found in cognitive psychology research literature is called the spotlight model. William James is credited with the spotlight model. He characterized visual attention to include focus, margin and fringe ranges. The focus area is where information from visual stimuli is acquired with the highest resolution. The center of the focus area, as calculated geometrically, is where vision is directed. Around the focus area is a region called the fringe where attention is at a much lower resolution. The fringe area will expand out to a specific area where vision is cut off. This boundary was named the margin.

The Zoom Lens Model

The other model of visual focus developed in the early 1800's is called the zoom lens hypothesis of visual attention. This theory describes the ability to resize the area of visual focus and works similar to the way a zoom lens on a camera works. The wider the area of focus the less efficient the processing of visual information becomes because this theory assumes that a person's attention processing resources are fixed. The wider the area of focus the slower the visual information will get processed due to attention resources getting distributed over a wider and larger area.

Arthur Davis

3 DIVIDED ATTENTION

Paying attention to multiple tasks at once is divided attention and occurs most of the time in everyone's life. It is a very important capability in order to carry out even the most rudimentary of tasks in our daily lives. In fact, you will find yourself, or anyone for that matter, hardly ever occupied with just one task. Having a conversation on a cell phone and driving, music playing and doing homework or watching television and cooking are all examples of divided attention. A good example of this is driving a car. You can't drive a car without paying attention to several things at once: your speed, road signs, other cars on the road, etc. Not paying attention to all of these may result in tickets, or worse, accidents.

Practicing Divided Attention

Recent scientific studies indicate that divided attention can improve with practice. Researchers studied response time and accuracy performance on subjects that were reading short stories while writing words dictated to them at the same time. At first their performance was very bad when trying to do both things at once; however, after practicing for five days per week for at least 85 sessions, their accuracy significantly improved while performing both tasks. It was hypothesized that some tasks, if controlled, can be carried out "automatically" with practice resulting in using fewer and fewer attention resources. Another key factor in someone's ability to perform well in situations requiring divided attention is their intelligence. The higher someone's intelligence the more capability that person has of sharing attention resources among two tasks at once and performing them well enough to accomplish both.

Divided Attention Theories

Over the years many theories about divided attention have emerged. In the early 1970's a researcher by the name of Daniel Kahneman theorized that people have a single "pool" of resources they use for focusing their attention and they can be divided at will among several tasks at once. However today, Kahneman's theory seems too simplistic. When tasks require using the same modality (modalities being visual, verbal or auditory), it is difficult to concentrate on both tasks at the same time. Thus, writing something down and listening to someone speak are of the same modality and the tasks are likely to interfere with each other when concentrating on both of them at once.

The Attentional Resource Theory

This specific modality theory was modeled by researchers in the latter part of the 1970's. The modality theory is more satisfactory in describing the process of dividing attention between simple tasks. However, a more modern theory called the attentional resource theory is much better at dealing with how attention is divided between more complex tasks. As recently as 2012 researchers demonstrated how "automatizing" a complex task frees up attentional resources that are limited in capacity.

However, there are many variables that affect our capacity to focus our attention and concentrate on many tasks at the same time. Some of these, among others, are how complex and difficult the tasks are to do, how skilled a person is in doing the tasks at hand and how anxious and/or aroused a person is while attempting to complete the tasks. Attention is now understood to be a multifaceted process and is much better understood with scientific advances in brain and cognitive research.

Bottom Up And Top Down Attention

There have been recent studies in how attention is directed according to certain properties of the stimuli itself. This so called bottom up attention process theory explains that much of our attention is directed toward objects in our environment because of movement or loud noise. Attention directed in this way is not voluntary or planned but happens whether we want it or not. Top down attention is controlled by a person's "executive" mental process faculties and is goals oriented. This is basically what you choose to pay attention to. There will be more about the mental executive in later chapters.

Arthur Davis

4 THE IMPACT OF LOAD ON ATTENTION PROCESSING

A theory that is increasingly relevant for today's attention issues is one characterized as the cognitive load theory. The theory hypothesizes that there are two processes that affect a person's attention capabilities. One is perception and the other is cognitive. The perceptual process is defined as the person's ability to perceive or ignore task and non task stimuli. Some research indicates that with a big "load" of stimuli to perceive and process, if the stimuli are task related, it is much easier to not pay attention or ignore the non task stimuli. On the other hand, if there are very few stimuli at hand, your mind will perceive the non task stimuli as well.

The Cognitive Process Of Attention

The cognitive process is how the mind processes stimuli after it is perceived. The results of studies on this attention process show that one's ability to process information once perceived decreases as a person gets older. So as one gets older if there is a large amount of stimuli perceived, a smaller and smaller amount of non task related stimuli gets processed by the brain.

The Morse Code Phenomenon

Some people can be trained to process multiple stimuli by automating complex tasks. As an example, some highly skilled Morse code operators can accurately record 100 percent of a Morse code message and simultaneously carry on an important conversation. They have automated the task so completely that detecting, receiving and transcribing a Morse code message is so "automatic" that it requires essentially no "attention" paid to the task to perform it successfully.

However, there is also a heavy cost when trying to pay attention to too many things at once. When your attention is divided among several task based stimuli, your performance in doing any of them can degrade dramatically when none of the tasks are automated as described in the Morse code example.

Heavy Multitasking

Think about the many things you attempt to do all at once at any given time of the day. Even right now you may be reading this book while also doing one or several other tasks. You could be having the radio playing, browsing your email, playing a computer game in another browser or checking your text messages on your Smartphone.

Attempting to do these things or similar tasks all at once will put you in a classification of what researchers call "heavy multitasker". You think you are pretty good at balancing your attention among all of these reasonably complex tasks and are "doing" them with adequate performance. Recent research into this is showing you may not be as "good" at multitasking as you think you are.

All along, most people held the belief that multitasking is a good technique to enhance personal productivity. At first it makes sense to think that doing several unrelated tasks at one time is going to result in more results, accomplished faster. However, this new research suggests that changing from one task to another actually decreases productivity. It turns out that heavy multitaskers have more difficulty ignoring distractions than when they focus their attention resources on one task. Additionally, cognitive processing is also impaired even when there are minimal distractions.

Arthur Davis

5 THE HEAVY COST OF MULTITASKING

Multitasking can mean more than doing multiple tasks at a time. It can also mean rapidly switching from one task to another over and over. It can mean rapidly doing one task after another successively.

In measuring the impact of multitasking on performance, psychologists asked subjects of a study to switch tasks and measured the amount of time that was lost when doing so. One test conducted by Stephen Monsell and Robert Rogers found that those participating in the study were much slower when they were required to switch tasks than when they repeated them.

In a subsequent study completed in the year 2001, researchers Jeffrey Evans and Joshua Rubinstein concluded that not only were there substantial amounts of time lost switching from one task to another but the amount of time lost increased as the tasks the participants were asked to perform became more complex.

Multitasking And Your Brain

There is a brain function commonly referred to as the "mental executive". The mental executive takes on the management of certain cognitive processes and decides what tasks will be performed, how they will get performed and in what order. Psychologists who have conducted research on how the executive process works in the brain hypothesize that the executive function in the brain is a two-stage process. The first stage they call goal shifting where the executive chooses which task to do among those available and the second stage is called role activation where the executive "shifts gears" from one set of "task based rules" to ones that are needed to perform the task the executive has chosen.

Changing these rules can take only tenths of a second, which seems miniscule, but this lost time can add up dramatically when repeatedly switching from one task to another. In the reality of daily living, it may not seem harmful when doing mundane tasks simultaneously such as cooking and watching television. But when someone is doing dangerous, complex tasks such as driving in traffic or tasks where productivity is important such as work-related tasks, these seemingly small amounts of lost time can turn out to be critical. There have been estimates of up to a 40 percent loss of productivity caused by "mental blocks" created when switching tasks.

Technology Overload

Recently psychology scientists have determined that dealing with email, phone calls, text messages and other incoming information made possible by technology can alter the way people behave and think. This technological information overload is degrading our ability to focus our attention.

We are born with a "primitive impulse" to immediately react to threats as well as opportunities. When a threat or opportunity presents itself, the hormone dopamine is secreted into our brain and this becomes addictive. Without this constant "thrill", we become bored with life.

The Addicted Heavy Multitasker

Multitasking becomes an additive habit as people seek out more and faster dopamine squirts with each new and exciting task to switch to. They convince themselves that this is good because they are "getting more things accomplished" by multitasking. However, the result of continued multitasking is actually more stress as multitaskers experience problems associated with difficulty focusing and ignoring irrelevant information.

More concerning is that recent research shows that this fractured cognition and difficulty focusing linger long after the multitasking has ended. So your brain is eventually trained to operate this way even away from your Smartphone and computer.

The problem is that now people are consuming almost three times the information as they were in 1960. Attention is constantly shifting from one "shiny new object" to the next. When people use computers in their work, some studies conducted on multitasking have shown that they change from one window to another or check their email almost 37 times per hour!

Arthur Davis

6 GET TO KNOW HOW YOUR MIND WORKS

Your mind truly is an amazing thing. First of all, it can store and retrieve vast amounts of information. It is almost like a computer in that it can analyze and solve complex problems. Your brain can refine billions of pieces of data and combine them into one single feeling or thought. However, just like your car, Smartphone or computer, you must learn how your brain works so that you can use it to your full advantage. There is one simple thing that if you know about it, you can leverage the astonishing power of your mind. Even if you do know about it, if you ignore this one fact about your mind, you are going to hinder the fabulous power your mind holds for you.

Here is the secret:

As brilliant as your mind really is, it is designed to focus on one thing at once. Correct! At first this may seem simplistic and counter what you believe and have read previously, but if you think about it, most profound truths you learn in life are simple and obvious. As sophisticated and complex as your mind is with all of its astonishing capabilities, it can only concentrate on one thing at a time. This may seem unbelievable to you at first but try this exercise and see if you are not convinced afterwards.

Think about swimming underwater in the ocean with a school of sharks. Imagine the huge fins and sharp teeth swimming up to you. Is this clear in your mind? Keeping these images and thoughts in your mind then think about what is in the trunk of your car. Try as hard as you can to think about BOTH of these at the same time. Impossible? You may be able to switch from one of these thoughts to the next but you can't hold both at once. This "limitation" can work to your advantage or disadvantage depending on how you use it.

Your Assembly Line Mind

To drive the point home, it may be useful to make an analogy between your mind and a manufacturing assembly line. Try to think about how an assembly line works in a manufacturing facility. It is probably designed to produce more than one product...but not at the same time. One week it may assemble Smartphones then maybe the next week MP3 players. What would happen to the productivity of the assembly line if management decided to make computers and wheelchairs at the same time by alternating between the two? Would this work or just create chaos? If even one finished product did come out, bets are it would be full of flaws and produced with a tremendous amount of wasted time and energy. Your mind works the same way: One thing at a time...or wasted time, energy and perhaps chaos. So if billion dollar manufacturing facilities can only make one thing at a time, why should you be able to complete more than one task at a time?

Yet most of us try to multitask complex tasks. Just as soon as we start our mind working on a complex task such as driving, we try to switch our concentration onto something else such as texting or talking on the phone. When working we think about what we have to do as soon as we get home. While at home we think about something we forgot to do at work. The phone rings and now we are talking to someone about something entirely different. It is the exceptional, or you could call them gifted, person who has trained themselves to focus on just one task at a time long enough to let the power of his or her mind work on it without interruption until completed successfully.

Arthur Davis

7 SEVEN WAYS TO IMPROVE YOUR FOCUS AND CONCENTRATION

One - Make A Conscious Decision To Concentrate

Forcefully Make A Decision To Concentrate

One of the cornerstones of focus and mental concentration is making the conscious decision that you are going to do it. You can't just expect it to happen magically. This may seem simplistic but without forcefully deciding to change the way we are thinking, it is easy to just move throughout the day from one unplanned experience to the next without any real concentration on anything.

Two - Get Away From Distractions

Be aware that wondrous information technology breakthroughs in recent years have made us much more knowledgeable, connected and productive in ways that people living in the 1940's and 1950's would marvel at...but at what price? The Smartphone, Facebook, Twitter, and countless other technological wonders in our lives are distracting us on a minute by minute basis. Possibly addicting us to that dopamine squirt we get when surfing from one thing to the next all day long.

Relentlessly train your mind to block these addictive distractions when you are concentrating on an important task that needs to get done! Stop this addicted cognitive overload by ignoring email, closing out your browsers, turning off your Smartphone and iPad and dealing with whatever minute by minute distractions not listed here but you know are present in your life.

Three - Know What Outcome You Want

Set Goals

Concentration is best achieved if you know when the task at hand is completed or at least when the time you have allotted to focus on the task will be over. A good technique to conquer big goals is to break them down into smaller achievable chunks. This way you can feel good about what you have achieved in smaller time frames and rack up several "victories" along the way: Like writing a novel 500 words at time.

Four – Consistency

Choose Your Space Wisely And Have A Regular Schedule

If you ever get to personally know successful and talented people, be they authors, musicians, executives or athletes who are at the top of their game, you will probably find out they practice and hone their skills purposefully at set times and in specific places. A consistent environment and schedule will help to make it easier to block distractions and get into a rhythm of focus and concentration. Choose a small number of places and times that will maximize your chances for concentration without interruption. These can be varied just enough so that your work will not get boring. Erratic or rapid changes will not serve to solidify the concentration habit you are trying to develop.

Five – Don't Overdo It

Especially when you first get started, your mind needs to "get into shape" similar to starting an exercise routine. During your preset concentration periods you should take small breaks and there should be long periods between sessions so as not to get overloaded or burned out. During your short breaks while in these concentration sessions, it is best NOT to direct your attention to some other task but to truly let your mental resources rest. In between scheduled concentration sessions, it is important to get enough sleep and proper nutrition.

Six – Don't Give Up. Practice Makes Perfect!

Don't expect to be able to do this overnight. Your multitasking, hurried life probably did not happen overnight and changing it will not happen overnight either. In order to get good at focus and concentration, you are going to have to practice it and get better at doing it with each and every attempt...just like any other life skill you set out to get. Before you get started, clear your mind and focus on having a fun and successful session.

Seven – Focus And Concentration Are Whole Body Experiences

Your thinking comes from your brain which is a body organ. It is going to function much better if you take care of it. You will be able to focus and concentrate at a much higher level with adequate sleep and nutrition. However, you will notice a marked decrease in your mental capacity if you abuse your body and mind with alcohol, mind altering drugs and excessive stress. Take care of your mind and body and it will take care to you!

Arthur Davis

CONCLUSION

Think about your life like you would a photograph. When it is out of focus what you see is blurry or fuzzy. The last time you took a picture and the camera focus was off or not working, were you disappointed with the result? Could be that the people or the scenery were so blurred that it made everything in the picture unrecognizable. What did you do with it? Did you delete it or toss it out if it was a print? The picture did not meet the quality standard you wanted when you snapped your camera to take the picture. Chalk it up as a learning experience. I guess you could blame it on the camera or the lighting or whatever.

Now take a snapshot of your life. Does it look and feel out of focus and blurry?

Well to fix this, the first thing you need is a clear image of the ideal life you have in mind for yourself. You must first create the desired picture in your imagination. All things in the "picture" must be in focus. This means clearly defined and envisioned.

Next, as you go about doing what is necessary to make this ideal picture a reality, you must remain focused and not let distractions in your surroundings take you off course and "blur your vision" of this picture.

This may seem too easy but it is totally amazing, especially with today's electronic gadgetry, how easily and fast people get distracted, off course and permanently out of focus. If you don't believe me then I suggest you ask someone you know well to describe a picture of their ideal life. I would bet that 90 percent of the time you will not get a clear concise description of it. It will be vague and imprecise. It will be full of generalities without concrete specifics. Sometimes you will get a description of what is not in the picture. This is no way to clearly achieve what someone wants.

So if your life is out of focus and you don't seem to be able to accomplish anything worthwhile, monitor what you are doing everyday all day long. Are you engaged in a series of junk emails, irrelevant tweets, mindless Facebook pages and computer games? Or, do you have a clear picture of what you want to accomplish and a roadmap you can follow to accomplish it? Next, what are you doing to get what needs to be done to make this picture come to life? Set goals. Schedule focus and concentration sessions and get away from ALL distractions. This means your Smartphone and browsers that are not related to the task at hand.

Good luck and try reading this book again. This time in a space and at a time where you can concentrate on it!

www.ingramcontent.com/pod-product-compliance
Lightning Source LLC
Chambersburg PA
CBHW070721180526
45167CB00004B/1564